WEATHER
ON EARTH

BY HILARY MAYBAUM

P9-EED-847

TABLE OF CONTENTS

4

6

16

How are the day-to-day
conditions of Earth's
atmosphere described,
measured, and predicted?

28

26

IT'S RAINING.

THE BRIDGE IN THE CLOUDS

You have probably watched clouds float by overhead. You may even have thought they resembled certain objects or characters. Perhaps you've flown through clouds on an airplane. But have you ever walked through a cloud? If you tread the sidewalks of the Golden Gate Bridge, the odds are you will!

The Golden Gate Bridge spans the San Francisco Bay in California. It is named after the entrance to the bay from the Pacific Ocean, the Golden Gate Strait. This steel bridge is over 2,700 meters (nearly 2 miles) long and 27 meters (90 feet) wide. It was completed in 1937, and for the next twenty-seven years it was the longest suspension bridge in the world.

The bridge is known for its beautiful art deco design and bright orange color. Its architect, Irving Morrow, selected a color called "International Orange." He wanted the bridge to blend well with its natural setting. Orange complements the warm colors of the land and contrasts with the cool colors of the sea and sky. It also provides good visibility for the ships that pass through the bay.

The Golden Gate Bridge is also famous for the thick blanket of fog that obscures the bridge during most of August. This weather phenomenon has been captured in countless photographs. Like all fog that rolls into low-lying areas, this famous fog is actually a kind of cloud.

Why do clouds form so low in Earth's atmosphere? Even more puzzling, what is it about the month of August that causes this cloud to hang around the Golden Gate Bridge?

Read this book to find the answer to this question—and to many other interesting questions about the weather in all its fascinating forms.

▲ San Francisco's Golden Gate Bridge is often draped in fog.

HEAT ENERGY

What will you wear to school tomorrow—a light jacket or a heavy coat? Will you need an umbrella? Should you bring your gloves, a hat, and maybe a scarf? The best way to make such decisions is to know about the weather. **Weather** is a set of atmospheric conditions at a given time.

Many elements make up weather. One element is air temperature—the coolness or warmness of the air. Another element is cloud cover, or how much of the sky is concealed by clouds. There are also rain, snow, and other forms of precipitation. Weather also includes humidity, air pressure, and wind. Almost all weather occurs in the lowest layer of the atmosphere, the **troposphere** (TROH-puh-sfeer).

HOW DOES HEAT ENERGY AFFECT EARTH'S ATMOSPHERE?

ESSENTIAL VOCABULARY

ELEMENTS OF WEATHER

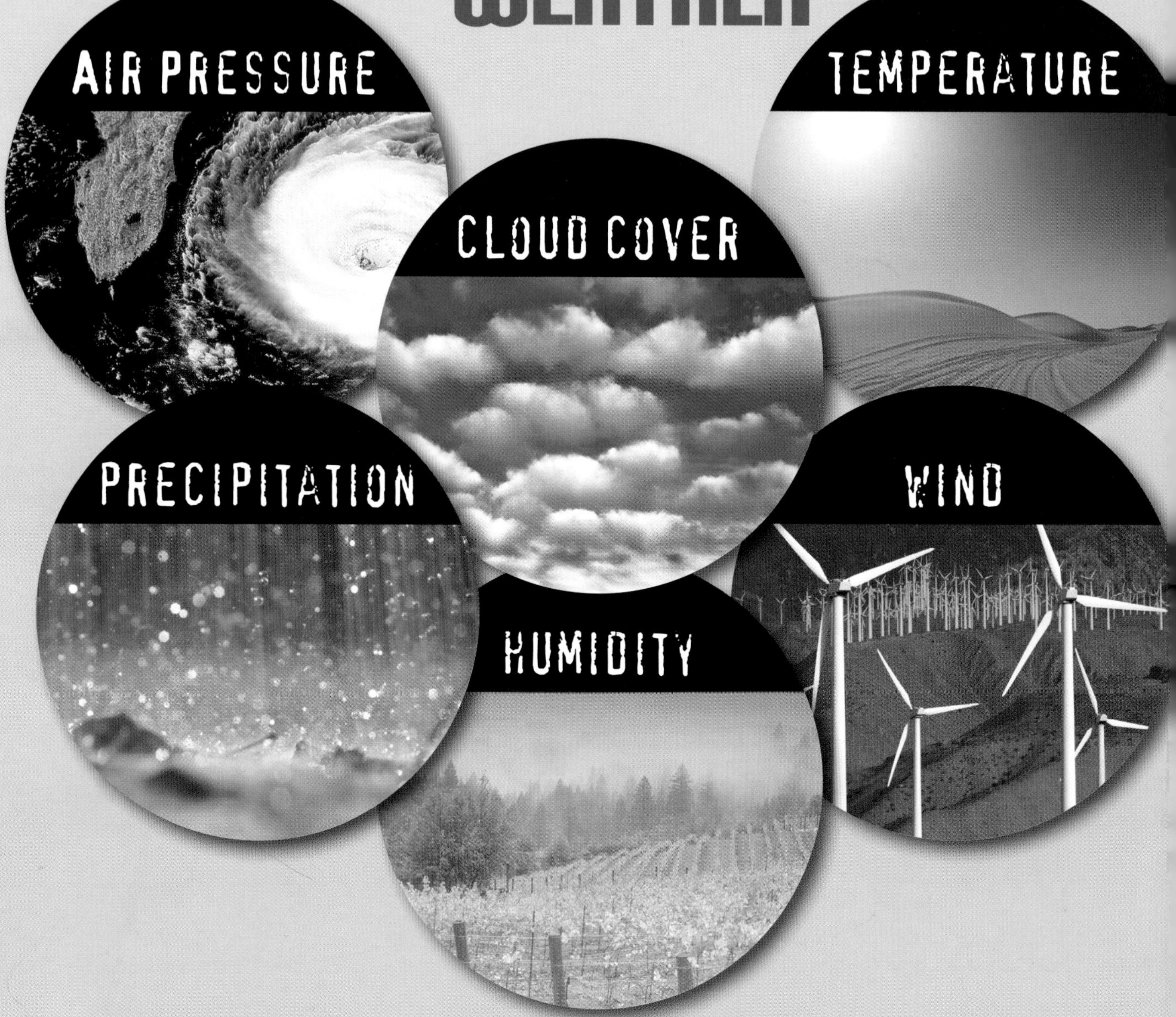

AIR PRESSURE

TEMPERATURE

CLOUD COVER

PRECIPITATION

WIND

HUMIDITY

EVERYDAY SCIENCE: WEATHER VERSUS CLIMATE

Climate and weather are not the same thing. Climate describes the average conditions of temperature and precipitation in a given region over a long period of time. Weather is the daily condition of Earth's atmosphere. These conditions change from day to day and even from moment to moment. For this reason, weather is less predictable than climate.

Energy from the Sun

Did you know that all weather is caused by the sun? All weather conditions need energy and they get this energy from the sun. The sun radiates, or gives off, solar energy in the form of electromagnetic waves. Most solar energy is in the form of light, some of which is visible and some of which is invisible. But all the light from the sun carries heat.

The light produced by the sun radiates through space to Earth in waves. These waves can be visible light, infrared light, ultraviolet light, microwaves, radio waves, radar, and x-rays.

Because light travels as waves, it does not need particles of matter to transmit it. Light can travel through a vacuum. The speed of light in a vacuum is nearly 300,000 kilometers (more than 186,000 miles) per second. That's *really* fast! When the sun's light is absorbed by your skin, you perceive it as heat.

Heat Transfer

The direct transfer of energy by electromagnetic waves is called **radiation**. If you've ever felt the heat of the sun on your face, you've experienced radiation. The same thing happens when the sun's radiation reaches Earth's surface. The land and water absorb energy through radiation.

Besides radiation, solar energy can also be transferred in other ways. Have you ever walked barefoot on black pavement after the sun has been shining on it? How does it feel? The pavement is hot because it has absorbed the sun's radiation. Your feet also feel the heat. This is because of a process called conduction. **Conduction** is the transfer of heat energy by substances that are in direct contact with one another. It takes place mostly in solids.

▲ Radiation transfers solar energy to Earth.

9

Heat is not easily conducted through gases. However, in the atmosphere, the air closest to Earth's surface is warmed by conduction. Only the air within a few meters of the surface can be heated this way. The rest of the troposphere is heated by convection.

Convection is the transfer of heat energy from one place to another in the form of currents. A current is the movement, or flow, of a liquid or a gas in one direction. The heated air molecules near the ground have more energy. As the heated particles move around and bump into each other, they move farther apart. This heated air becomes less dense than the cooler air above it. This makes the cooler, denser air sink and the warmer, less dense air rise. The upward and downward movement creates convection currents. These currents move heat throughout the troposphere. In all cases of heat transfer, heat energy moves from a warmer object or place to a cooler object or place.

CHECKPOINT

TALK IT OVER
Think of some other examples of heat transfer in your daily life. Share your examples with a classmate. Classify each example by the type of heat transfer: conduction, convection, radiation, or a combination of these.

METHODS OF HEAT TRANSFER

RADIATION
The transfer of heat energy in the form of waves.

CONDUCTION
The transfer of heat energy by direct contact of the particles of matter.

CONVECTION
The transfer of heat energy through currents in liquids or gases.

The Greenhouse Effect

The **greenhouse effect** is a naturally occurring process that contributes to the heating of Earth's surface and atmosphere. Without the greenhouse effect, the average temperature of Earth would be much colder than it is now. Life as we know it would not exist.

The greenhouse effect is the trapping of heat energy by clouds and gases in the atmosphere. Solar energy enters Earth's atmosphere and gets held inside. The greenhouse effect involves the three different methods of heat transfer. First, solar energy reaches Earth's atmosphere by radiation. Clouds, particles, and certain gases in the atmosphere reflect some of this energy back into space and also absorb some of the energy. The remaining energy passes through the atmosphere to Earth's surface. Here, some of it is reflected back into space, and some is absorbed and converted into heat.

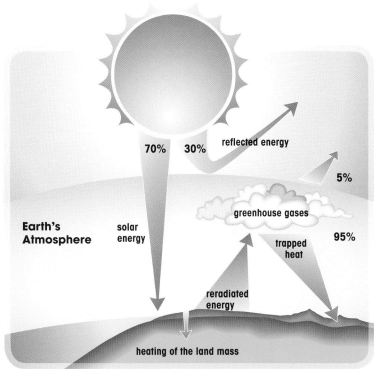

▲ Of all the heat transferred to Earth's atmosphere and surface, almost half is reflected back out to space. Substances in the atmosphere trap some of this reflected heat, keeping Earth warm. This trapping of heat by atmospheric gases is called the greenhouse effect.

Some of the heat is transmitted to the lower atmosphere by conduction and convection. Some of it is reradiated in the direction of space. However, not much of this heat escapes Earth. Certain atmospheric gases—such as carbon dioxide, water vapor, and methane—trap more than 90 percent of Earth's reradiated energy. These "greenhouse gases" then radiate energy back to Earth's surface, where it is absorbed, and the cycle continues. The trapping of heat energy by greenhouse gases causes additional warming of Earth's surface and atmosphere.

EVERYDAY SCIENCE: THE SUN AND YOU

Solar energy keeps Earth warm, but not all of it is helpful. Some of the energy the sun produces is in the form of ultraviolet, or UV, light. UV light causes harmful changes in skin cells. In some cases, these changes can cause skin cancer. Sunscreen and proper clothing can help protect skin from harmful UV light.

Variations in Heat Energy

The amount of solar energy that reaches Earth's surface is not the same everywhere on Earth. The energy is greatest at the equator, where the amount of direct sunlight is greatest most of the year. This is because the sun's rays hit the equator at angles between 66 ½ degrees and 90 degrees. These angles concentrate the sun's rays in a small area. The result is that areas at or near the equator receive the greatest amount of heat energy.

Areas north or south of the equator receive less solar energy because the sun's rays are less direct most of the year. This means the sunlight is spread out over a wider area and cannot warm Earth as much. The result is that areas farther from the equator—and closer to the poles—receive the least amount of heat energy. The differences in heat energy from north to south across Earth's surface greatly influence weather.

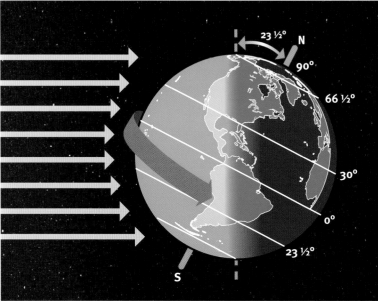

▲ The sun warms Earth unevenly. The sun's rays are most direct at the equator. Places farther from the equator receive less heat energy, because the same amount of sunlight is spread across a larger area.

SCIENCE AND MATH: CONVERTING TEMPERATURE UNITS

Celsius is the International System unit, or SI unit. This means that most countries around the world use the Celsius scale (°C) to measure temperature. The United States and a few other countries use the Fahrenheit scale (°F). Mathematical formulas are used to convert one scale to another.

To convert a temperature in °C to °F, use this formula: $F = \left(C \times \frac{9}{5} \right) + 32$

To convert a temperature in °F to °C, use this formula: $C = \left(F - 32 \right) \times \frac{5}{9}$

Measuring Temperature

You may describe the air as cold or warm, but just how cold or warm is it? To be accurate, you can measure the air's temperature. **Temperature** is a measurement of the average energy that particles of matter have. Temperature is directly related to heat energy. The more heat energy particles have, the faster they will move and the higher their temperature will be.

Two common temperature scales are the Fahrenheit (FAIR-en-hite) and Celsius (SEL-see-us) scales. Their units are expressed as degrees Fahrenheit (°F) and degrees Celsius (°C). On the Fahrenheit scale, water freezes at 32°F and boils at 212°F. On the Celsius scale, the freezing point of water is 0°C, and the boiling point is 100°C.

Temperature is measured with a **thermometer**. The liquid inside a thermometer will expand or contract depending on the level of heat energy. The liquids most frequently used are mercury and alcohol. When a liquid is heated, it expands and its volume increases. When a liquid is cooled, it contracts and its volume decreases. That is how a thermometer works.

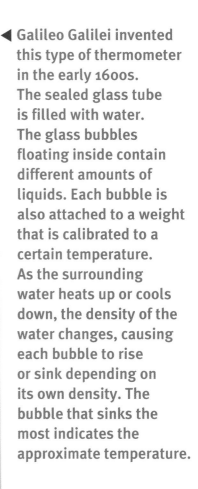

◀ Galileo Galilei invented this type of thermometer in the early 1600s. The sealed glass tube is filled with water. The glass bubbles floating inside contain different amounts of liquids. Each bubble is also attached to a weight that is calibrated to a certain temperature. As the surrounding water heats up or cools down, the density of the water changes, causing each bubble to rise or sink depending on its own density. The bubble that sinks the most indicates the approximate temperature.

◀ This thermometer uses a glass bulb filled with mercury. As the temperature rises, the mercury expands and moves up the tube. As the temperature falls, the mercury contracts and moves down the tube.

HANDS-ON SCIENCE: BUILD A BULB THERMOMETER

How does a liquid bulb thermometer work? Find out with this experiment.

TIME REQUIRED

45 minutes

GROUP SIZE

small groups of four

MATERIALS NEEDED

- narrow-necked glass bottle
- water
- food coloring
- modeling clay
- clear drinking straw
- index card
- tape
- marking pen
- shallow pan of warm water
- ice cubes

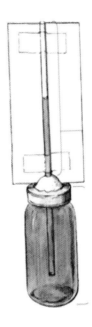

PROCEDURE

1. Fill the bottle to the top with water. Add several drops of food coloring. Gently swirl the bottle to mix the contents.

2. Shape the clay around the straw, about 5 centimeters (2 inches) from the end.

3. Insert the straw into the bottle. Mold the clay to form an airtight seal. Do not let the straw touch the bottom of the bottle.

4. Push the narrow edge of an index card into the clay behind the straw. Tape the straw to the index card.

5. The water will rise above the clay seal. Use a marking pen to mark the water level on the index card with the number 1. This is the level at room temperature.

6. Set the bottle in a pan of warm water. Wait for the water level to stop rising. Mark the new level with the number 2.

7. Fill the pan of water with ice cubes. Wait for the water level to stop falling. Mark the new level with the number 3.

ANALYSIS

- What happened to the water in the straw when it was heated? Why?

- What happened to the water in the straw when it was cooled? Why?

- How does this experiment model a liquid bulb thermometer, such as an alcohol or mercury thermometer?

SUMMING UP

- Weather is a set of constantly changing conditions of the atmosphere.
- The elements of weather are temperature, cloud cover, precipitation, humidity, air pressure, and wind.
- Weather is greatly influenced by the sun's energy, which reaches Earth by radiation.
- Once absorbed by Earth's surface, the energy is transferred to the atmosphere by conduction, convection, and radiation.
- The atmosphere traps a great deal of this heat energy as a result of the greenhouse effect.
- The amount of heat energy that areas on Earth receive depends on their distance north or south of the equator.
- The most direct rays of the sun strike Earth at the equator, and the least direct strike Earth at the poles.
- A thermometer measures heat energy as temperature.

PUTTING IT ALL TOGETHER

Choose one of the activities below. Work independently, in pairs, or in a small group. Share your responses with the class. Listen to other groups present their responses.

1. You read on page 9 that conduction is the transfer of heat energy from one object to another through direct contact. Find a wall or other surface that has been absorbing the sun's energy. First, place the palm of your hand against your arm. Note how cool or warm your palm feels on your arm. Next, place the same palm against the warmed surface you found. Keep your palm on that surface until it feels warmer than before. Then take your palm away and place it on your arm again. How does its temperature compare to before? Explain what happened to your palm and your arm in terms of conduction.

2. Based on what you read on page 12, where might you expect to find the coldest temperatures on Earth's surface? Where might you expect to find the hottest temperatures? Find out where the coldest and hottest temperatures on the surface of Earth were measured. How do these places compare to your predictions? Explain why there are differences.

3. Use an outdoor thermometer to collect temperature data. Measure how the outside air temperature changes during the day. Collect measurements every hour for five to seven hours. Record your measurements in a data table, making sure to note the units of measurement. Then organize your data in a chart or graph. Explain how the temperature changed during the day. What was the highest temperature? What was the lowest temperature? What factors do you think might have affected your temperature readings?

AIR PRESSURE AND HUMIDITY

SCIENCE AND MATH: CALCULATING DENSITY

You can calculate density by dividing the mass of an object by its volume. This calculation can be stated as a mathematical formula: $D = M/V$, where D = density, M = mass, and V = volume. Some common units of density are kilograms per cubic meter (kg/m^3), grams per cubic centimeter (g/cm^3), and grams per milliliter (g/mL).

HOW ARE AIR PRESSURE AND HUMIDITY RELATED TO WEATHER?

Earth's atmosphere is like a blanket that is hundreds of kilometers thick. Right now, that entire blanket of air is pressing down on you! Do you feel it? Most likely, you don't.

Air is matter. It is made up of particles too small to be seen with the unaided eye. But air, like all forms of matter, has weight. In fact, one liter (about one quart) of air weighs about 1.3 grams (less than one-tenth of an ounce). Air presses down on Earth's surface with all the force of its weight. The **air pressure** at any point on Earth's surface is equal to the weight of the air particles above that point. You are not aware of air pressure because it is nearly equal from all directions—including from inside you pushing out.

Air pressure is not the same at every point on Earth's surface. Why? Air pressure depends on the density of the air. **Density** measures how much matter is in a given volume of an object. Density is equal to mass per unit of volume. Denser air has more mass per unit of volume than does less dense air. This means that denser air exerts more air pressure than does less dense air. The density of air, and thus air pressure, depends on three factors: temperature, elevation, and water vapor.

Factors That Affect Air Pressure

Temperature

The particles that make up air are in constant motion. As air warms, these particles move faster and farther apart. The decrease in the number of particles in the same space, or volume, results in a decrease in density. Less dense air exerts less air pressure. The opposite effect happens when air cools. The particles move more slowly and closer together. The increase results in greater density. More dense air exerts more pressure.

▼ Air pressure decreases as elevation increases. The air is thinner, or less dense, at higher elevations. This often makes breathing difficult.

Elevation

Elevation, or altitude, is height above sea level. As elevation increases, the air becomes thinner, or less dense. Thus air pressure decreases with increasing elevation. Air pressure increases with decreasing elevation.

Water Vapor

About 99 percent of the air is made up of the gases nitrogen and oxygen. The remaining 1 percent of the air is water vapor, the gaseous form of water. Water vapor particles have less mass than particle of nitrogen or oxygen, so the more water vapor in the air, the less dense the air is. This means that air with a large amount of water vapor in it exerts less pressure than drier air.

CHECKPOINT

Reread
Reread page 18 to review the factors that affect air pressure. Relate each factor to density. Discuss the factors with a partner to check your understanding.

Measuring Air Pressure

Another term for air pressure is barometric pressure. A **barometer** is an instrument that measures air pressure. All barometers contain some kind of fluid that responds to changing pressure. That fluid can be water, air, or mercury. Because air pressure varies with temperature and elevation, standard air pressure is measured at 0°C at sea level. There are two types of barometers: the mercury barometer and the aneroid (A-nuh-roid) barometer.

A barometer can help you predict changes in the weather. Remember, water vapor affects air pressure. If the barometer reading goes down, it means there is more water vapor in the air. A rainstorm is probably on its way. A rising barometer means there is less water vapor in the air. You can expect clear skies ahead.

- An aneroid barometer consists of an airtight metal capsule from which most of the air has been removed. The capsule has flexible sides that respond to changes in air pressure.

- As changes in air pressure alter the thickness of the capsule, a pointer moves along a dial. The pointer indicates the new air pressure.

- Air pressure can be read in millibars, millimeters of mercury, or inches of mercury. The average air pressure at sea level is 1,013.2 millibars, or 29.9 inches of mercury.

▼ The French inventor Lucien Vidie developed the aneroid barometer in 1843.

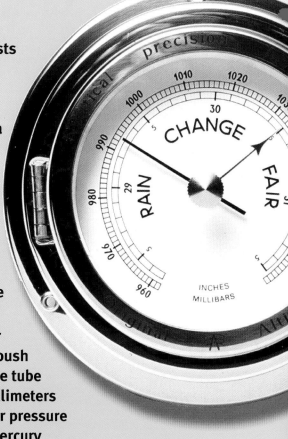

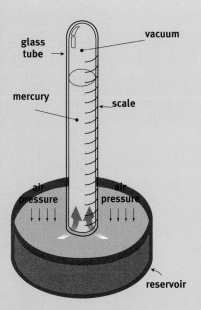

▲ The Italian scientist Evangelista Torricelli invented the first mercury barometer in 1643.

- A mercury barometer consists of a glass tube about 900 millimeters (36 inches) high that is filled with mercury. The tube rests in a container of mercury, open end down.

- The barometer measures air pressure by balancing the weight of the column of mercury in the glass tube against the weight of the air pushing down on the mercury in the container.

- At sea level and 0°C, the air exerts enough pressure to push the column of mercury in the tube to a height of about 760 millimeters (nearly 30 inches). When air pressure decreases, the column of mercury falls. When air pressure increases, the column of mercury rises.

Humidity

Have you ever been outdoors on a damp day? How did the air feel? Damp air contains a lot of water vapor. The amount of water vapor in the air is called **humidity**. The water vapor comes from evaporation—an important process in the water cycle. During evaporation, liquid water from many sources on Earth turns into water vapor. The reason why humid weather can be uncomfortable is because a person's perspiration cannot readily evaporate in humid air, since there is already so much water vapor.

Absolute and Relative Humidity

The amount of water vapor in a given volume of air is called absolute humidity. By itself, this number is not very useful. That's because humidity depends on the air temperature. Warm air can hold more water vapor than cold air.

▼ You cannot see water vapor, but it is all around you. At low elevations, the air can get very thick with water vapor. The extra humidity can cause fog to form near the ground.

Relative humidity is expressed as the percentage of water vapor the air is holding compared to the amount it could hold at a particular temperature. Relative humidity can be used to judge how comfortable the air feels. Most people are comfortable when relative humidity is 50 to 70 percent. On muggy or damp days, it may be as high as 80 to 90 percent. When the relative humidity is 100 percent, the air is said to be saturated. It is holding all the water vapor it can possibly hold at the given temperature.

Measuring Humidity

A **hygrometer** (hy-GRAH-meh-ter) is a device for measuring humidity. The simplest kind is made up of two thermometers. One thermometer has a dry bulb. It measures the current air temperature. The bulb of the other thermometer is wrapped in wet fabric and measures the air temperature as a result of evaporation. The absolute humidity of the air affects how quickly water evaporates from the wet fabric. The relative humidity is determined by comparing the readings from the two thermometers.

When the difference between the dry-bulb and wet-bulb thermometer readings is large, relative humidity is low. When the difference is small, relative humidity is high.

EVERYDAY SCIENCE: BAD HAIR DAYS

Like the air, human hair absorbs moisture. Hair lengthens when it is wet and shortens as it dries. This natural reaction makes human hair a pretty good indicator of humidity. High humidity can cause frizzy hair or weigh the hair down. In fact, in high humidity, hair length can change by 3 percent!

▼ Warm, moist air may get pushed up the slope of a mountain. As the rising air cools, it condenses, forming a cloud above the mountaintop.

Clouds

No matter what they look like, all clouds have the same ingredients—water droplets and ice crystals. The droplets float in the air, as tiny as specks of flour.

Some clouds form by convection. Sunlight warms Earth's surface. The ground heats the air above it, making the air less dense. The warm air rises and cools. Because cool air cannot hold as much water vapor as warm air, the rising air soon reaches its dew point. The **dew point** is the temperature at which the air is saturated. At dew point, the water vapor condenses, or changes into a liquid. In the air, condensation needs a "seed"—a tiny, solid particle such as a piece of dust that acts as a surface. The water vapor in the air condenses on this surface as liquid droplets.

Clouds can also form if air is forced upward. When moving air meets a hill or mountain, it moves up the slope. Moist air condenses as it rises and cools. This is why you often see clouds above mountaintops.

▼ Water vapor in the air can condense on hard surfaces, such as grass. Here, condensation forms dew. When water vapor condenses on tiny specks of dust in the air, it can form clouds.

Types of Clouds

Clouds are classified by their appearance and by the altitude of the base of the cloud. They can be grouped into four categories according to appearance.

- Cirrus (SEER-us) clouds are high clouds that are wispy and hairlike.
- Cumulus (KYOO-myuh-lus) clouds are white clouds that are puffy or lumpy.
- Nimbus (NIM-bus) clouds are dark clouds that are puffy or lumpy.
- Stratus (STRAY-tus) clouds are sheets of low-lying clouds that spread across the sky.

The prefixes *cirro-*, *alto-*, and *strato-* can be added to the name of the cloud type. The prefix indicates the cloud's altitude. There are ten different cloud types. Each type of cloud is associated with certain types of weather.

Cirro-
High clouds above 6,000 meters (20,000 feet)

Alto-
Clouds at 2,000 to 6,000 meters (6,500 to 20,000 feet)

Strato-
Clouds below 2,000 meters (6,500 feet)

1. CIRRUS

Cirrus clouds are the highest clouds in the sky, forming at altitudes above 6 kilometers. They appear thin and feathery and are sometimes called mares' tails. Cirrus clouds are made of ice crystals. They accompany fair weather, but they often indicate a change to rain or snow within twenty-four hours.

4. ALTOSTRATUS

The prefix *alto-* means "high." An altostratus cloud usually covers the whole sky and has a gray or blue-gray appearance. The sun or moon may shine through, but will appear blurred.

6. STRATOCUMULUS

The prefix *strato-* refers to low-altitude clouds. Stratocumulus clouds are small, puffy, gray clouds. They often form in rows with blue sky visible between them. Although rain rarely occurs with these clouds, they can turn into a type associated with continuous rain or snow.

8. NIMBOSTRATUS

Nimbostratus clouds bring rain and snow. They appear as a layer of dark, wet-looking clouds.

2. CIRROSTRATUS

These sheetlike thin clouds usually cover the entire sky.

3. CIRROCUMULUS

The prefix *cirro-* refers to high-altitude clouds (12–18 kilometers in altitude). Cirrocumulus clouds are high, puffy clouds that appear in long rows and with ripples. Usually seen in the winter, they indicate fair but cold weather. In tropical regions, they may signal an approaching hurricane.

5. ALTOCUMULUS

These are cumulus clouds at altitudes of between 2 and 5 kilometers. They are white to gray in color and may have dark sides. These clouds are often associated with rain, snow, or the arrival of cold air.

7. STRATUS

Stratus is a Latin word that means "spreading out." Stratus clouds are low, flat, dull-gray clouds that sometimes seem to cover the entire sky. Stratus clouds resemble fog that does not touch the ground. These clouds are formed by the slow rising of air. Stratus clouds are associated with light mist, drizzle, or rain.

9. CUMULUS

The word *cumulus* is Latin for "heap" or "pile." Puffy and white, cumulus clouds look like piles of floating cotton balls. Plain cumulus clouds mean good weather, and are often called "fair-weather clouds." But cumulus clouds can quickly change into other types.

10. CUMULONIMBUS

The suffix *-nimbus* indicates a cloud that produces precipitation. If cumulonimbus clouds are on the horizon, rain is soon likely. If they grow larger and darker, a thunderstorm with strong winds can be expected. These clouds are also associated with snow, hail, and tornadoes.

23

Precipitation

Any form of water that falls from the atmosphere to Earth is called **precipitation**. The most common forms of precipitation are rain, snow, sleet, freezing rain, and hail. Like evaporation and condensation, precipitation is an essential part of the water cycle. Water that evaporates from lakes, oceans, rivers, and ponds forms water vapor in the air. Water vapor then condenses back into liquid water, forming clouds. When the water droplets in the clouds get heavy enough, they fall to the ground as precipitation.

CHECKPOINT

Visualize It
How might you use common household materials to make a rain gauge? Sketch your design on a blank sheet of paper.

rain gauge   tipping bucket

RAIN

SNOW

FREEZING RAIN

HAIL

SLEET

▲ When it comes to precipitation, most people automatically think of rain. However, precipitation has many other forms.

Measuring Precipitation

Precipitation in the form of rain is measured with an instrument called a rain gauge. The simplest rain gauge is shaped like a funneled cylinder and has measuring marks in millimeters, inches, or both. A tipping bucket is a kind of rain gauge with at least one small container that can swing freely. When precipitation fills the container, it tips over. At each tip, an electronic device records the amount of precipitation.

SUMMING UP

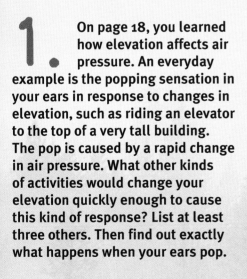

- The air particles that make up Earth's atmosphere press down on Earth's surface.
- Air pressure is the weight of all the air particles above a particular point.
- Air pressure depends on density.
- The density of air, in turn, depends on temperature, elevation, and humidity.
- Humidity is the amount of water vapor in the air.
- A barometer can be used to measure air pressure. A hygrometer measures humidity.
- Water vapor can condense in the air to form clouds.
- All the water comes back down to Earth's surface by precipitation.
- A rain gauge is an instrument used to measure precipitation.

PUTTING IT ALL TOGETHER

Choose one of the activities below. Work independently, in pairs, or in a small group. Share your responses with the class. Listen to other groups present their responses.

1. On page 18, you learned how elevation affects air pressure. An everyday example is the popping sensation in your ears in response to changes in elevation, such as riding an elevator to the top of a very tall building. The pop is caused by a rapid change in air pressure. What other kinds of activities would change your elevation quickly enough to cause this kind of response? List at least three others. Then find out exactly what happens when your ears pop.

2. Choose a type of cloud and write a short story from the cloud's point of view. Title your story, "A Day in the Life of a _____ Cloud." Your story could describe how and where the cloud formed, how it moved, or how it changed. Be sure to describe the weather conditions on that particular day.

3. A proverb is a short saying that expresses a common or useful thought. Choose one of the following weather proverbs to research. Find out what the proverb means and whether it has a scientific basis. Then determine whether the proverb is valid.

Rain before seven, clear before eleven.

Birds flying low, expect rain and a blow.

Red sky at night: sailors' delight. Red sky at morning: sailors take warning.

When dew is on the grass, rain will never come to pass.

BUDDING METEOROLOGIST

CARTOONIST'S NOTEBOOK ○ ILLUSTRATED BY PETE PACHOUMIS

MOM SAID TO BRING YOU SOME FOOD AND THIS THING OF YOURS. SHE SAID YOU'VE BEEN IN HERE ALL DAY.

STILL WANT TO BE A METEOROLOGIST, HUH?

I'M PREPARING MY WEATHER REPORT FOR A CLASS PRESENTATION.

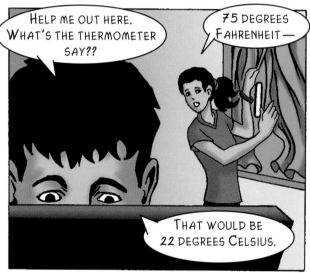

HELP ME OUT HERE. WHAT'S THE THERMOMETER SAY??

75 DEGREES FAHRENHEIT—

THAT WOULD BE 22 DEGREES CELSIUS.

NOW THE BAROMETERS.

28 ON THE MERCURY BAROMETER AND 990 ON THE OTHER ONE.

THAT ONE IS CALLED AN ANEROID BAROMETER. BOTH MEASURE AIR PRESSURE...

HISS ES A HY GROM—

SWALLOW FIRST, THEN TALK.

This is a hygrometer. It measures humidity. 71...69.

Aha! Now look at this. It's just as I thought. A warm front is coming up from the south and cool air is coming off the ocean.

If you say so.

Yep.

The temperature is rising and the barometer is falling, which means there is more water vapor in the air. Plus, the hygrometer shows little difference, which means relative humidity is high.

All together, that can mean only one thing—

It's RAINING!

Exactly! Hey! How did YOU know?

Technology can help us measure and observe the world around us, but we still must rely on our five senses.

What are some types of observations that only people can make?

27

AIR in MOTION

You have now learned about five of the elements that make up weather: air temperature, air pressure, humidity, cloud cover, and precipitation. Perhaps you are now asking yourself why weather changes. The answer to that question involves the movements of huge masses of air. It also involves the remaining element of weather—wind.

Air Masses

An **air mass** is a large dome of air with similar temperature and humidity in each of its layers. A typical air mass is thousands of kilometers wide. Air masses move because changes in temperature cause changes in density and air pressure. The movements are based on the basic principle you have been learning about: warm air rises and cool air sinks.

Air masses are named according to where they form. This location is called a source region. It is usually a large area, such as an ocean or continent. A source region gives an air mass its two important characteristics: temperature and humidity. Once an air mass moves out of its source region, it can change depending on the conditions around it.

Air masses are named using symbols that indicate temperature and humidity. Continental air masses form over land. Maritime air masses form over oceans. Depending on their combinations of temperature and humidity, six types of air masses can form. Five of the six are found over North America.

CHECKPOINT

Make Connections
Use the map to identify the air mass closest to where you live. Based on what you have read, write a sentence or two describing that air mass. Be sure to include its name, temperature, and humidity.

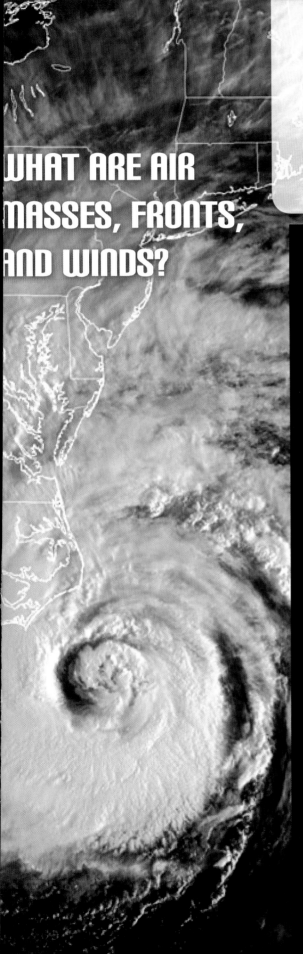

WHAT ARE AIR MASSES, FRONTS, AND WINDS?

AIR MASSES IN NORTH AMERICA

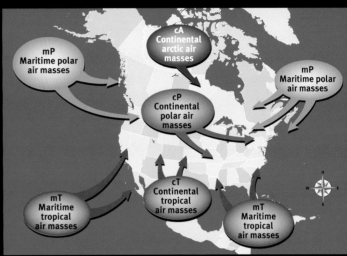

Humidity Symbols		
Symbol	Name	Humidity
c	Continental	dry
m	Maritime	wet

Temperature Symbols		
Symbol	Name	Temperature
A or AA	Arctic or Antarctic	cold
P	Polar	cool
T	Tropical	warm
E	Equatorial	hot

An uppercase letter tells the temperature of the air mass. A lower-case letter tells the humidity. Five of the six types of air masses form over North America. Notice the absence of an equatorial (E) air mass.

cA Continental arctic air masses form over the frigid regions of Northern Canada. They are very cold and dry.

mP Maritime polar air masses are cool and moist. They bring mild weather to coastal locations.

cP Continental polar air masses originate over the northern plains of Canada. They carry cold and dry air southward.

cT Continental tropical air masses form over hot, dry land areas such as deserts. They are warm and very dry.

mT Maritime tropical air masses develop over subtropical oceans. They carry heat and moisture northward to the United States.

Fronts

When two different air masses come together, they do not mix readily. Instead, a **front**, or boundary, forms between the air masses. Fronts usually result in clouds and precipitation. In fact, people in many places of the world rely on fronts to bring needed rainfall. The weather conditions at a front depend on the conditions of the colliding air masses. The diagrams show the four types of fronts that can form.

WARM AND COLD WEATHER FRONTS

▲ A long band of clouds can form when a warm front collides with a cold front.

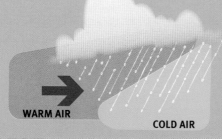

WARM FRONT

When a warm, moist air mass pushes into a cooler air mass, the warm air slides up and over the cooler air. As the warm air rises along the gentle slope between the two air masses, it cools. Its water vapor condenses to form clouds and rain. The rain band may be a few hundred kilometers across. It is often followed by mild, humid air. Warm fronts are slow-moving compared to cold fronts.

COLD FRONT

If a cold, dry air mass meets a warmer air mass, the cold air wedges under the warm air. The cold air pushes the warm air upward along a steep slope. Along this slope, a band of tall cumulus clouds often forms. The clouds produce brief, heavy showers and thunderstorms. Clear, blue skies and cooler temperatures usually follow Cold fronts move twice as fast as warm fronts.

STATIONARY FRONT

Sometimes when two air masses meet, neither one moves. Both remain in place, or stationary. Clouds build up and may last for some time. Precipitation may be off-and-on while the front remains.

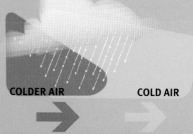

OCCLUDED FRONT

Cold fronts travel faster than warmer fronts. A cold front can also catch up with another cold front and push up underneath it, forcing the warmer air up. This pushing is known as an occlusion (uh-KLOO-zhun).

HANDS-ON SCIENCE: MODEL A WEATHER FRONT

When two different air masses come together, rather than mixing with each other, they collide. A front forms between the air masses. What happens when a warm air mass collides with a cold air mass? You will model this event in this experiment.

TIME REQUIRED
20 minutes

GROUP SIZE
small groups of four

MATERIALS NEEDED

- scissors
- heavy cardboard
- rectangular clear-glass cooking dish
- food coloring
- pitcher of water
- stirrer
- cooking oil

PROCEDURE

1. Cut out a cardboard barrier to fit the glass dish.
2. Place the barrier vertically in the center of the dish to separate the two sides. Make the seal as tight as possible, but do not tape it.
3. Add several drops of food coloring to the water in the pitcher and stir.
4. Pour the cooking oil into one side of the dish.
5. Pour the colored water into the other side of the dish.
6. Carefully lift the barrier. Observe what happens.

ANALYSIS

- What did the cooking oil represent? What did the colored water represent?
- Draw a diagram of your observations. What happened when the barrier was lifted?
- How does this activity model a weather front?

Winds

How would you describe the wind? A brisk slap? A steady push? A gentle caress? Wind can be all of these. **Wind** is simply air in motion.

The source of wind is solar energy. By day, the sun heats Earth's surface. Convection transfers this heat to the air above the surface, warming the air.

Recall that warm air rises because it is less dense than cool air. As it rises, the air cools. Cool air is denser than warm air so it sinks.

In the atmosphere, air constantly circulates between cool, upper regions and warmer, lower regions. These up-and-down air movements form pressure systems. A high-pressure system, or a high, forms where air is moving downward in the atmosphere. A low-pressure system, or a low, forms where air is rising.

Air flows naturally from areas of higher pressure to areas of lower pressure. The resulting winds can blow on a large, global scale or on a much smaller, local scale.

SCIENCE AND TECHNOLOGY: WIND ENERGY

Have you ever flown a kite? If so, you have used wind energy. A wind turbine is a device for gathering wind energy and converting it to mechanical energy. This energy can then be used for specific tasks, such as grinding grain or pumping water. It can also be supplied to a generator, which converts it into electricity. A large group of wind turbines, a so-called wind farm, can produce enough electrical energy to power a small town.

This wind farm can make enough clean energy to run a small town.

Sea Breezes and Land Breezes

Although both land and water may get the same amount of sunlight, land heats more quickly. Warm air over the land rises first, forming a low. Cooler air moves in from the water to replace it. This pattern of convection is called a sea breeze. At night, the pattern reverses. Land cools more quickly than water. The breeze flows from the land toward the water, forming a land breeze.

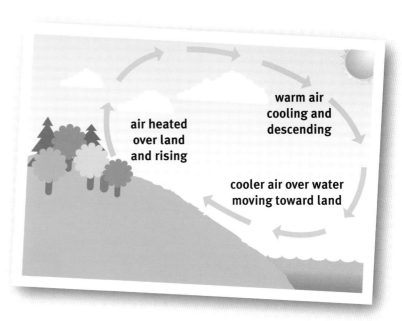

air heated over land and rising

warm air cooling and descending

cooler air over water moving toward land

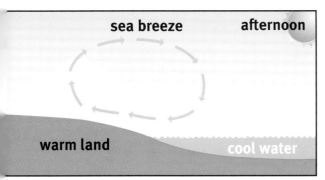

sea breeze   afternoon

warm land   cool water

SEA BREEZE—Day
- The land warms faster than the water.
- Air heated over the land rises.
- The air cools and sinks.
- Cooler air moves toward land.

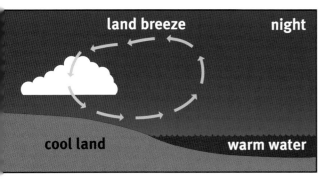

land breeze   night

cool land   warm water

LAND BREEZE—Night
- The land cools faster than the water.
- Warmed air over the water rises.
- The air cools and sinks.
- Cooler air moves toward the water

Global Winds

The movement of air from highs to lows also creates winds that travel the globe. These global winds, or trade winds, move warm air from the equator toward the poles. They also move cold air from the poles toward the equator. If Earth did not spin on its axis, global winds would move straight north and south. Instead, they move diagonally across the globe.

When winds blow over water, friction drags the water in the same direction. The movement of water in one direction creates a flow called a current. Ocean currents—like global winds—transport heat from warmer regions to colder regions.

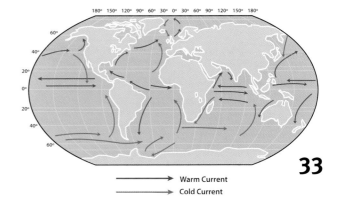

Warm Current
Cold Current

33

Measuring Winds

When wind is measured, two characteristics are described: speed and direction. Wind speed is measured with an **anemometer** (a-neh-MAH-meh-ter). This device has three cups shaped like spheres. When the cups catch the wind, they rotate around a vertical axis. The anemometer counts how many times these cups pass a fixed point. The speed is an average measurement taken over a period of time—usually a few minutes. It is frequently expressed as meters per second or miles per hour.

Often, an anemometer will have a wind vane attached to it. A wind vane indicates the direction from which the wind blows. The vane is designed to point into the flow of the wind. Winds are named for the direction from which they blow. If the arrow of a wind vane points to the north, the wind is coming from the north and is called a northerly wind.

A wind vane points in the ▶ direction from which the wind blows. The wind in this picture is from the south.

{ **The Root of the Meaning:**

ANEMOMETER

The prefix anemo- comes from the Latin word *animus*, meaning "breath." The suffix -meter refers to any device used for measuring. It comes from the Greek word *metron*, which means "to measure."

▲ An anemometer measures the speed of the wind.

SUMMING UP

- The moving air that circulates in Earth's atmosphere takes the form of air masses with similar temperature and humidity throughout.
- Air masses can form over land or water and can have high or low air pressure.
- Areas of high air pressure ("highs") often have light winds and clear skies.
- Low-pressure areas ("lows") usually have cloudy, rainy skies.
- When different air masses collide, a front forms, and a change in the weather can occur.
- Wind is air in motion, the source of which is solar energy.
- Air is constantly circulating between cool, upper regions of the atmosphere and warmer, lower regions.
- These up-and-down air movements form pressure systems.
- Horizontal movements between land and sea form sea breezes and land breezes.
- The movement of air from highs to lows also creates winds that travel the globe. Global winds and ocean currents carry heat from the equator to the poles.

PUTTING IT ALL TOGETHER

Choose one of the activities below. Work independently, in pairs, or in a small group. Share your responses with the class. Listen to other groups present their responses.

1. Examine a weather map from a newspaper or online weather report. Identify one of the air masses. It may be marked with an H for "high" or an L for "low." Describe the air mass in terms of temperature and humidity, and indicate the direction in which it is traveling. Write your description in a paragraph. Reread pages 28–29 to infer the source of the air mass you selected, such as cA or mP. Explain your reasoning.

2. Reread page 30. In your own words, explain what happens when a cold front forms. How is it different from the formation of a warm front? Find out which types of clouds can form from a warm front and a cold front. Organize your findings in a chart or table.

3. Page 33 discusses different types of winds. One interesting type of wind is the jet stream—a fast-moving band of air currents found in Earth's upper atmosphere. The location of the jet stream is very important to airline pilots. Find out more about the jet stream and its importance in flight. Prepare a report of your findings. Illustrate your report with pictures, such as photographs or diagrams.

PREDICTING THE WEATHER

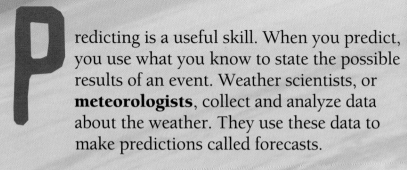

Predicting is a useful skill. When you predict, you use what you know to state the possible results of an event. Weather scientists, or **meteorologists**, collect and analyze data about the weather. They use these data to make predictions called forecasts.

Gathering Weather Data

To make good predictions, scientists need accurate information. Several times each day, weather observers record the conditions throughout the troposphere, the lowest layer of the atmosphere.

Weather Stations

A weather station is a place with many different instruments to measure the elements of weather. A typical weather station has all the instruments you've read about in this book. These include thermometers, barometers, hygrometers, rain gauges, anemometers, and wind vanes.

There are about 10,000 weather stations around the world. Several thousand more are on ships anchored in different parts of the ocean. Together, these weather stations form a global network. Many send their data automatically to places called forecasting centers.

WHAT METHODS ARE USED TO PREDICT THE WEATHER?

This weather station in Antarctica uses a wind vane, among other tools, to measure conditions.

CHECKPOINT

Read More About It
Read more about weather stations at the library or on a trusted Internet site, such as the home page of the National Weather Service. Find out how many weather stations are in your state.

SCIENCE AND HISTORY: RADAR

Radar stands for *ra*dio *d*etection *an*d *r*anging. This is a technology that can find objects from far distances. It was first developed before World War II. Early radar used radio waves to detect warcraft. Meteorologists later adopted this technology to map the location and strength of rainfall.

Weather Balloons

At some weather stations, **weather balloons** are released each day at the same time. The balloons travel into the middle atmosphere, or stratosphere. When a balloon reaches a high elevation, it bursts and falls to the ground. The distance the balloon travels and where the balloon falls indicate the speed and direction of the wind. Some weather balloons have long cables attached to instruments that measure temperature, air pressure, and humidity. When the balloon bursts, these instruments come down on a parachute.

Weather Satellites and Aviation Weather Tools

Weather satellites travel high above Earth. They orbit the upper atmosphere, or exosphere. They take detailed pictures and collect temperature data. Some satellites orbit Earth at the same speed at which Earth spins on its axis. This way, the satellite always stays over the same point. Others move between the North and South poles.

Many commercial aircraft carry instruments that collect, record, and transmit weather data. These instruments measure pressure, temperature, wind speed, wind direction, and humidity. Radar is used to see where precipitation is occurring.

Analyzing and Reporting Data

Supercomputers are used to analyze weather data. These extremely fast and powerful computers can make millions of calculations each second. They plot weather observations on a map of the world. The map is divided into square grids. Each grid has several boxes stacked above it. This way, scientists can see all the conditions from Earth's surface upward.

Forecasting

The weather forecasters you see on television or hear on the radio do not work alone. As you've read, weather systems extend hundreds of kilometers. It may take a team of thousands of meteorologists in many countries to infer how these different systems interact, with the end result of one weather forecast. The forecast is shown on a weather map with special symbols. Radio and e-mail messages provide instant updates.

CAREERS IN SCIENCE: METEOROLOGIST

A meteorologist uses technology to observe, understand, explain, and predict the weather. This person can have many different jobs: daily forecasting, researching, teaching, broadcasting, or consulting. To be a meteorologist, you usually need a college degree. If you enjoy physical science, Earth science, or mathematics, this might be the career for you!

In the past, forecasters used only pencils, paper, and calculators. Supercomputers are a big improvement over the tools of the past. Today's computer models take into account all of the changing conditions in the atmosphere. Still, weather data are too complicated to accurately predict weather more than a few days in advance.

Meteorology offers the challenge of forecasting a natural event and the satisfaction of seeing how that forecast affects the lives of thousands of people.

Weather Maps

Meteorologists use the data they compile from satellite photographs, radar images, or computer models to create weather maps. Weather maps are graphic representations that show the current weather in a given region. Meteorologists use these maps to track weather patterns and make predictions, or forecast, the future weather.

The most common weather map you will come across is called a surface analysis. These maps tend to show major pressure systems and precipitation. We often see these maps on television or in newspapers. Each line and symbol represents a condition in the atmosphere. Other weather maps give even more detail about temperature, precipitation, and winds.

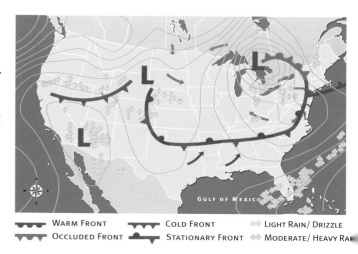

WARM FRONT	COLD FRONT	LIGHT RAIN/ DRIZZLE
OCCLUDED FRONT	STATIONARY FRONT	MODERATE/ HEAVY RAI

▲ **This weather map is called a surface analysis. It shows the presence of highs and lows in the United States on July 2. It also shows several different fronts and forms of precipitation. You can see three low-pressure systems over the United States.**

SUMMING UP

- Every day, meteorologists collect information about the conditions of the atmosphere from thousands of weather stations around the world —on land and at sea.
- Weather balloons, radar, satellites, and aircraft help gather data from high altitudes.
- The data is shared by forecasting centers in different countries.
- Teams of scientists use supercomputers and weather models to analyze the data, plot it on maps, and make forecasts.
- The forecasts are shown on maps and are constantly updated.

PUTTING IT ALL TOGETHER

Choose one of the activities below. Work independently, in pairs, or in a small group. Share your responses with the class. Listen to other groups present their responses.

1. Review the sources of weather data you read about on pages 38–39. Choose one of these methods and find out more about it. Determine the advantages and disadvantages of the method you selected. Collect pictures and make diagrams to explain in detail how the method works. Present your findings in a report. Try to use pictures, sounds, or video in your report.

2. Set up your own weather station. First, determine the data that you will collect and the instruments that you will need. Find out if you can make any of the instruments and, if so, how. Figure out how you will analyze and report the data. Then put your ideas into a written plan. Make sure to get the approval of your teacher, parent or guardian, and your school's administration. Once your plan is approved, see it through.

3. Interview a meteorologist. First, compose a list of at least ten questions to ask. For instance, what does the job require? Where and when does this person work? What kind of training did he or she receive? With the help of a teacher, parent, or guardian, contact a local television or radio station, university, or museum to find out if you could talk with a meteorologist who works there. Then request permission to contact that person by phone, e-mail, or in person. Make an appointment and keep it. Write up your questions and answers in a report. Be sure to send a thank-you note to the person you interviewed.

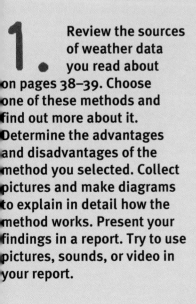

SUMMER FOG

Now that you've read this book, you know that fog is a form of condensation—a low-lying stratus cloud. In technical terms, fog is defined as a visibility of less than 1 kilometer (3,300 feet). To form, this ground-hugging cloud needs the same conditions required for any cloud. These include the cooling of moist air and condensing of microscopic water droplets in the atmosphere.

Summers in San Francisco are usually overcast and cool. The fog that settles over the San Francisco Bay in early morning often burns off by afternoon. Why, then, does the fog that blankets the Golden Gate Bridge persist for long parts of the day throughout August? The answer is an ocean current.

The California Current is a flow of cold water that runs along the coast of California. It is part of a larger current called the North Pacific Gyre. The California Current begins near the southwestern tip of Canada and travels south through the Pacific Ocean, ending at Baja, California.

August in San Francisco is a warm month. The air temperature averages 23°C (73°F) in the daytime. When the cold air over the California Current mixes with the warm August air surrounding San Francisco Bay, the warm air cools. Its water vapor condenses, forming a thick, white fog. Fueled by the continuous cool air from the California Current, this fog moves over the Golden Gate Strait and settles on the Golden Gate Bridge.

When cool air from the California Current mixes with warm air over the San Francisco Bay, a dense fog forms. This fog settles on the Golden Gate Bridge.

How to Write a Weather Report

Y ou can write about the weather and forecast, or predict, the weather in the future. All you have to do is follow these steps and you can create your own weather report.

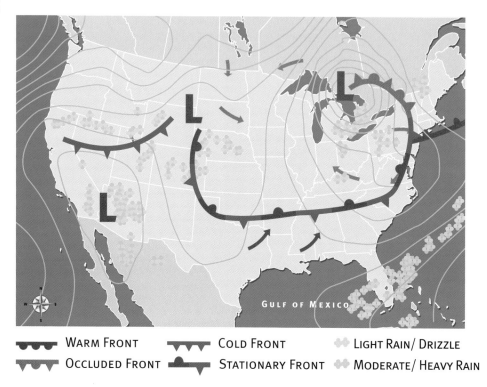

| WARM FRONT | COLD FRONT | LIGHT RAIN/ DRIZZLE |
| OCCLUDED FRONT | STATIONARY FRONT | MODERATE/ HEAVY RAIN |

GULF OF MEXICO

STEP 1
Choose an area of interest to you. Then use a home, school, or local library online connection to visit the National Weather Service Web site. Click on the main page map for the weather in the area you want to write about. You can also use other online weather Web sites and printed newspaper reports to support your research.

STEP 2
Start with existing conditions. Begin your report with the current weather for whatever area you are researching. Look at the current weather data for temperature, humidity, air pressure, precipitation, cloud cover, and wind. Be sure to state the known facts before making any predictions.

STEP 3
Read related eyewitness accounts to see if they can add to your report. Also mention weather alerts, or unusual weather events, if any.

STEP 4
Next, move on to the forecast. You can look at the current conditions and make predictions based on some of the information you learned in this book. You can also verify these predictions with the 5-day forecasts you see online or in print. Remember that while the current conditions are factual, and can be definite statements, your predictions about what will happen tomorrow or the next day must be phrased to reflect uncertainty. Choose qualifying words and phrases like "may," "expected," and "forecasted" instead of "will." The weather is never guaranteed!

Sample Weather Report

Central Arizona, July 2

Today's high temperature was 43 degrees Celsius (110 degrees Fahrenheit). The low temperature in some parts dipped down as far as 34 degrees Celsius (94 degrees Fahrenheit). The relative humidity was 75%. Scattered thunderstorms developed this afternoon across central Arizona as a result of an incoming cold front from the North. Winds blew at the surface mostly out of the north and east.

A strong storm moved across the northern and central sections, then dropped southeastward. This storm produced a lot of lightning, as well as small hail and heavy rainfall. Some minor flooding was reported in Flagstaff, with several observers claiming up to an inch of rain in about one hour!

Farther south, the storm brought some tree branches down in Payson. Luckily, no damage or injuries occurred. Rainfall across the southern part of the state was less eventful, with most areas staying completely dry but cloudy. The "big winner" for this event was of course Flagstaff, which measured a record-breaking 0.95 inches of rain today!

Tomorrow it looks like you may be able to put away your umbrellas. With this cold front passing, clear blue skies should be on the way tomorrow. More comfortable temperatures are expected in the low- to mid-nineties.

air mass	(AIR MAS) *noun* a large dome of air with similar temperature and humidity within each of its layers (page 28)
air pressure	(AIR PREH-sher) *noun* the weight of air particles pressing down on Earth's surface at any point (page 17)
anemometer	(a-neh-MAH-meh-ter) *noun* a device for measuring wind speed (page 34)
barometer	(buh-RAH-meh-ter) *noun* an instrument that measures air pressure (page 19)
conduction	(kun-DUK-shun) *noun* the transfer of heat energy by direct contact of particles of matter (page 9)
convection	(kun-VEK-shun) *noun* the transfer of heat energy from one place to another in the form of currents (page 10)
density	(DEN-sih-tee) *noun* how much matter an object has; the amount of mass per unit of volume (page 17)
dew point	(DOO POINT) *noun* the temperature at which the air is saturated with water vapor (page 21)
front	(FRUNT) *noun* a boundary between two different air masses (page 30)
greenhouse effect	(GREEN-hows ih-FEKT) *noun* the trapping of heat energy by gases and clouds in the atmosphere (page 11)